NIST Technical Note TN 1614

Alarm Information for Decision Support

William D. Davis
Stephen Kerber
Adam Barowy
Engineering Laboratory
Fire Research Division
National Institute of Standards and Technology
Gaithersburg, MD 20899-8664

November, 2010

U.S. Department of Commerce
Gary Locke, Secretary

National Institute of Standards and Technology
Patrick D. Gallagher, Director

Certain commercial entities, equipment, or materials may be identified in this document in order to describe an experimental procedure or concept adequately. Such identification is not intended to imply recommendation or endorsement by the National Institute of Standards and Technology, nor is it intended to imply that the entities, materials, or equipment are necessarily the best available for the purpose.

Alarm Information for Decision Support

William D. Davis, Stephen Kerber, Adam Barowy

Contents

Table of Figures iii

Abstract iv

Introduction 1

Building 2

Experimental Procedure 5

Fire Fighter Needs 6

Decision Support Systems 7

    Fire Room 3 8

    Fire Room 4 17

Discussion 23

Uncertainty Analysis 25

Conclusion 25

References 26

Table of Figures

Figure 1. Grey shaded areas are portions of the school used for the experiments. Thermocouple arrays are indicated in red by the designation TC. 4

Figure 2. Fuel package for classroom fires showing the wood pallets, excelsior, foam mats, and ignition point. The second ignition point is on the side opposite the ignition point shown in the figure. 6

Figure 3. Modeling layout for the fire tests. Dashed lines are virtual boundaries used to simulate the corridor as a series of rooms. 9

Figure 4. Fire Room 3. The squares are the prediction of the average layer temperature using SDFM where SDFM1 designates the fire room predictions. The rest of the plots are thermocouple data at the designated location below the ceiling. 11

Figure 5. Hallway labeled Room 2. The squares are the prediction of the average layer temperature using SDFM where SDFM2 designates room 2 predictions. The rest of the plots are thermocouple data at the designated location below the ceiling. 12

Figure 6. Hallway labeled Room 4. The squares are the prediction of the average layer temperature using SDFM where SDFM4 designates room 4 predictions. The rest of the plots are thermocouple data at the designated location below the ceiling. 13

Figure 7. Hallway labeled Room 5. The squares are the prediction of the average layer temperature using SDFM where SDFM5 designates predictions in room 5. The rest of the plots are thermocouple data at the designated location below the ceiling. 14

Figure 8. Hallway labeled room 8. The squares are the prediction of the average layer temperature using SDFM where SDFM8 designates predictions in room 8. The rest of the plots are thermocouple data at the designated location below the ceiling. 15

Figure 9. Layer Height measured from the floor as predicted by SDFM. 16

Figure 10. Fire Room 4. The squares are the prediction of the average layer temperature using SDFM where SDFM1 stands for fire room. The rest of the plots are thermocouple data at the designated location below the ceiling. 19

Figure 11. Hallway labeled Room 4. The squares are the prediction of the average layer temperature using SDFM where SDFM4 stands for room 4 prediction. The rest of the plots are thermocouple data at the designated location below the ceiling. 20

Figure 12. Hallway labeled Room 5. The squares are the prediction of the average layer temperature using SDFM where SDFM5 stands for room 5 prediction. The rest of the plots are thermocouple data at the designated location below the ceiling. 21

Figure 13. Hallway labeled Room 7. The squares are the prediction of the average layer temperature using SDFM where SDFM7 and SDFM8 stand for predictions for room 7 and 8. The rest of the plots are thermocouple data at the designated location below the ceiling. 22

Figure 14. Layer Height measured from the floor as predicted by SDFM. 23

Abstract

Experiments conducted in a Toledo, Ohio school are used to examine the use of signals or data from simulated heat detectors by incident command. Two approaches are examined. The first is to use the signal from ceiling-mounted thermocouples to deduce conditions in the fire room and hallway. The second approach is to process these data using the Sensor-Driven Fire Model to provide additional information to incident command. It is found that both methods show promise, but additional testing is necessary before either method is deployed.

Key Words: computer modeling, fire experiments, heat detectors, decision support, heat alarms

Introduction

The capability of building systems to provide real-time information to incident commanders has evolved from concept to standards development in the course of less than a decade[1]. As the ability of building systems to deliver this information evolves, an analysis of the capability of a building alarm system used for tactical decisions needs to be investigated. In this paper, the information that could be provided by heat alarms is examined using full-scale room fires in a school. This information is compared with the capabilities of fire fighter turn-out gear in order to compare fire fighter needs and limitations with available information.

Historically, heat alarms for fire systems in buildings are designed to provide an alarm signal when a specific set-point for a detector is reached. If the analog or digital signal produced by the heat alarm is monitored, then a temperature/time profile for that particular alarm can be developed. To make use of this temperature/time profile, key temperatures must be identified that can be used for tactical decisions by an incident commander. These key temperatures can be associated with performance limits of fire fighter protective clothing, can be used to identify fire spread or can be a measure of structural integrity.

Basic fire types that can be encountered in a structure include flaming fires, smoldering fires, fires in walls, attic fires, and explosions that result in fire. Flaming fires will develop a hot plume and ceiling jet which is just the extension of the plume running below the ceiling in the room of origin. Given sufficient time, a hot smoke layer will develop in the room and will also spread to fill all connected rooms with hot smoke. Fires in walls may be sufficiently concealed or have not grown enough to penetrate through walls so that only smoke will begin to fill the room. This type of fire may resemble a smoldering fire in a room in that both types of fires produce smoke but very little heat. Attic fires represent a fire type where the fire is located above the alarm unless there is an alarm located in the attic. In this case, the alarm is not immersed in the hot gases until the fire has broken through and below the ceiling. Fires that start with explosive events may have significant alarm failure such that the fire alarm monitoring may depend only on alarms located at the periphery of the explosive zone. The fires examined in this study are flaming fires that grow rapidly in a room where the heat alarms or sensors (in this case thermocouples) are immersed in the ceiling jet. The information content of remote alarms in corridors connected to the fire room is examined and compared with predictions using the Sensor-Driven Fire Model (SDFM)[2].

The primary purpose of the experiments described in this paper was to investigate the impact of ventilation on the smoke and fire from an emergency responder standpoint.[3] The long corridors in the building and the choice of instrumentation for the experiments provided an excellent opportunity to test the value of using heat alarm signals as a decision support tool for emergency response and to provide some validation of the SDFM.

A demonstration of the use of real-time signals from alarms for emergency responders can be found in a paper by Davis, et. al.[4] The demonstration was conducted in Wilson, NC on September 22, 2007 and was supported by the Wilson Fire Department, the staff at the Wilson Memorial Hospital, and researchers from Honeywell[*]. A virtual fire was initiated in a third floor wing of the Wilson Memorial Hospital and members of the Wilson Fire Department evaluated the En Route and On Site display information sent from the building to the fire truck, the mobile command center, the responding fire station, and the 911 center. The real-time information from the building was superimposed on the building floor plan using laptop computers. Based on the response from this demonstration, the National Electrical Manufacturers Association (NEMA) has begun the regulatory process to add an enroute display to its NEMA SB30 Fire Service and Annunciator and Interface standard, which is currently cited in National Fire Protection Association (NFPA) 72 Annex F, 2007.

Building

The experiments were conducted in a two-story 28,000 m^2 (300,000 ft^2) former high school. The school was originally constructed in 1956 and was added on to substantially until 1988. The structure is irregularly shaped with numerous sections and court yards but overall has the dimensions of 210 m (700 ft) wide by 130 m (425 ft) deep by 9 m (30 ft) tall (Figure 1). The building was constructed of masonry walls and steel column grids. The roof and floor systems were mostly steel deck on steel joists with reinforced concrete. Thermocouple arrays, consisting of ten thermocouples each, were located in the fire rooms and hallways and provided measurements of the changing environment. The top thermocouple in each array can also be regarded as a heat alarm, because the response of thermocouples to the changing environment is similar to heat alarms. The thermocouples were bare-bead, type K, with a 0.5 mm (0.02 in) nominal diameter. Each thermocouple array is designated by a letter associated with a location shown in Figure 1 and then a position. For example, the array in Fire Room 3 is presented in Table 1 and has a position uncertainty of 6 %^{2}.

---

[*] Certain commercial companies are identified in this paper in order to adequately specify the experimental procedure. Such identification does not imply recommendation or endorsement by the National Institute of Standards and Technology.

Table 1 Thermocouple array for fire room 3.

Thermocouple designation	Location Below Ceiling (m)	Location Below Ceiling (ft)
H_ceiling	0.03	.08 (1 in)
H_01	0.1	.25 (3 in)
H_02	0.2	.50 (6 in)
H_03	0.3	1.0
H_06	0.6	2.0
H_09	0.9	3.0
H_12	1.2	4.0
H_15	1.5	5.0
H_18	1.8	6.0
H_21	2.1	7.0

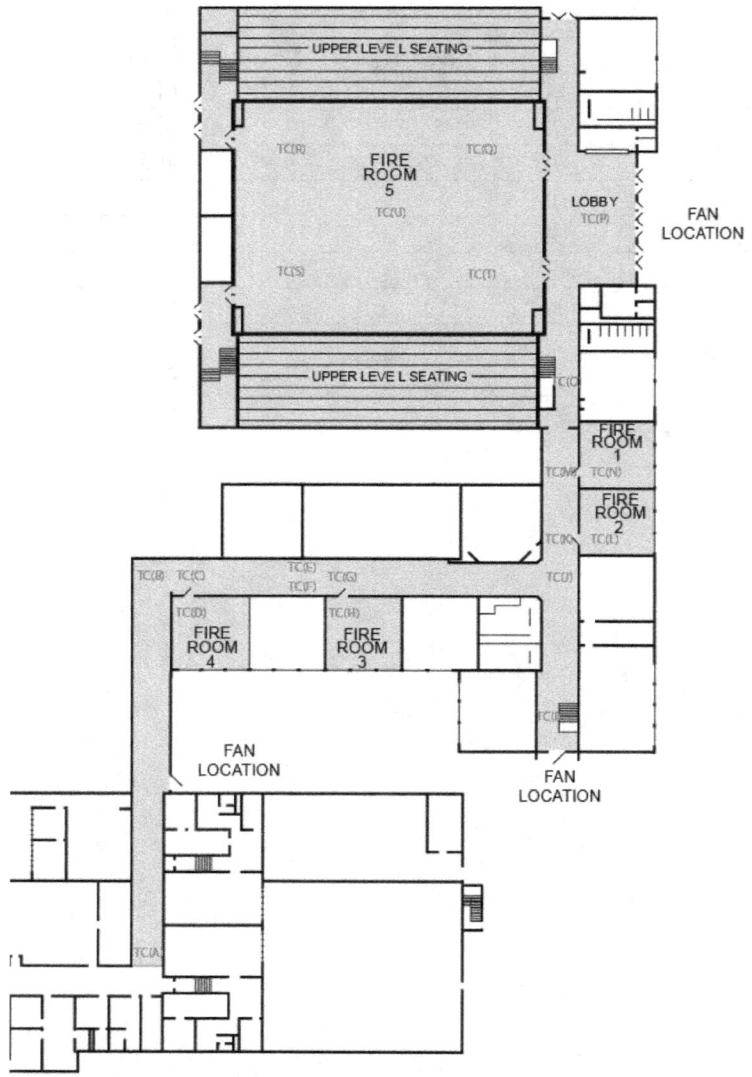

Figure 1. Grey shaded areas are portions of the school used for the experiments. Thermocouple arrays are indicated in red by the designation TC.

Experimental Procedure

Due to the complex floor plan and condition of the structure, a section of the building was isolated for the experiments. This section was to the right of the front entrance and included long stretches of hallway, numerous classrooms and a large volume gymnasium (Figure 1). The detailed floor plan in Figure 1 shows the experimental volume used during the experiments in light grey. Walls were constructed around the three stairwells in the hallway to eliminate smoke flow to the second floor and to better define the experimental volume for analysis. The two rooms labeled fire room 3 and fire room 4 were used for the comparisons in this paper. Each fire room was 8.28 m by 7.72 m with a 2.57 m ceiling height. The hallway connecting the rooms was 48.00 m long and 3.33 m wide with a 2.60 m high ceiling. The doorways between the rooms and hall were 0.89 m wide and 2.52 m high.

Each experiment began with all of the ventilation openings, doors and windows closed, with the exception of the door from the fire room to the hallway. The fuel package was ignited and the fire was allowed to grow for a minimum of six minutes. This was approximately the time when the fire reached its peak and became ventilation limited. See Table 2 for a detailed time history for each experiment.

Table 2 Timeline of Experiments for Fire Rooms 3 and 4

Fire Room 3		Fire Room 4	
Time (s)	Event	Time (s)	Event
0	Background	0	Background
60	Ignition	120	Ignition
420	Hallway Door 1 Open	480	Hallway Door 1 Open
560	SVU On	600	0.69 m (27 in.) Fan On
560	Long Ramp Up	734	0.69 m (27 in.) Fan Off/Door Closed
750	SVU Off/Door Closed	821	FR 4 Window Open
810	SVU On	881	0.69 m (27 in.) Fan On
1140	SVU Off	1165	Water Applied
1168	Water On	1300	End of Test
1300	End of Test		
SVU = Special Ventilation Unit			

The fuel load for the classroom fire experiments consisted of three components. The main component was wood pallets. The wood pallets were stacked flat. The second component was excelsior or shredded wood to allow the electric match to ignite the pallets. Each experiment utilized 8 kg of excelsior layered between the pallets. The final component was foam mats. The foam mats measured 1.5 m x 0.6 m x 0.05 m (4.8 ft x 2 ft x 0.2 ft). Each fuel package was positioned in the center of the room and was remotely ignited with electric matches positioned on two sides of the pallets at the open ends (Figure 2).

Figure 2. Fuel package for classroom fires showing the wood pallets, excelsior, foam mats, and ignition point. The second ignition point is on the side opposite the ignition point shown in the figure.

The analysis in this paper will only include the time between ignition and the start the external ventilation which occurred at 560 s for fire room 3 and 600 s for fire room 4. Additional detail for the experimental setup is available[2].

Fire Fighter Needs

Fire fighters responding to a building fire need to know the fire location, fire size, how fast the fire is growing, the best way to approach the fire, and if sprinklers have activated. The fire location is relatively easy to deduce as it will be either at or near the location of the first alarm provided that there are a sufficient number of alarms in the building to provide adequate spatial resolution. Information from heat alarms can provide excellent guidance for locating the fire. Temperatures in excess of 500 °C (930 °F) are locations where flashover is possible[5,6] even if there is no fire presently at that location. If a heat alarm is registering a temperature in excess of 800 °C (1500 °F), flames are likely present at the alarm's location.[7]

Whereas ceiling temperatures can provide information concerning the presence of a fire in a room, temperatures below the ceiling where fire fighters work are equally important. Fire fighters and their equipment can survive high temperature regions for only limited times before needing to retreat and cool off. This is particularly true when fire fighters are engaged in search and rescue and do not have a hose stream to control excess temperatures. A report examining Personal Alert Safety Systems (PASS) device capabilities developed a set of thermal classes (Table 3) which are useful for evaluating environments where electronic gear can operate and defining the environment that fire fighters can withstand in full turn-out gear and equipped with breathing apparatus[8]. If a ceiling mounted heat detector is measuring a temperature that is consistent with one of the thermal classes, then there is a probability that the fire fighter may be exposed to these conditions although the temperature at the level of the fire fighter may be lower than at the ceiling due to the two layer nature of a developing fire.

Table 3 Thermal Classes [Ref 5]

Thermal Class	Maximum Time (min)	Maximum Temperature (°C)/(°F)	Maximum Flux (kW/m^2)
I	25	100/212	1
II	15	160/320	2
III	5	260/500	10
IV	<1	>260/500	>10
V	<<1	>500/932	Flashover

Other temperatures that are of interest are associated with structural issues such as concrete and steel failure, concrete spalling, and wood and steel truss failure. The thermal environment also affects the tactical decisions for search and rescue efforts. For structural issues, the duration of exposure is almost as important as the temperature since concrete and steel must absorb significant amounts of heat to reach failure modes.

Decision Support Systems

Decision support systems (DSS) use building information from building sensors and systems to provide information support for tactical decisions. While a DSS can make use of a variety of sources, the two DSS investigated in this paper use temperature sensing. The information content and accuracy of the two different DSS are analyzed. The first system uses the temperatures of thermocouples mounted close to the ceiling which would mimic ceiling mounted heat alarm (CMHA) installations. The observed or reported temperatures are then used to evaluate conditions in the building for emergency operations. These conditions are presented in tabular form for each fire experiment

The second system, the Sensor-Driven Fire Model (SDFM)[2], also uses the temperatures from the top thermocouples in the fire room but processes them to provide additional information about the fire environment. It does not use the information from the thermal arrays in the hallways except as a measure of the reliability of its projections for the building environment. The fire location and size, depth and temperature of the smoke layer, and fire spread are calculated and decision support information concerning hazards, particularly visibility, temperature hazards, and flashover is generated by the model.

The systems involved operate in real-time and require only existing hardware to implement. The SDFM can be operated using a standard desk-top PC.

For this set of experiments, the default positioning algorithm was used to identify the location of the heat alarm. The SDFM was set to identify a small (4 kW) source which meant that the model issued an alarm almost as soon as the fuel source was ignited. The SDFM provided the following warnings as shown in Table 4.

Table 4 SDFM Warnings

Warning	Criteria
Visibility limited (VL)	Bottom of smoke layer 2 m above the floor
Toxic/thermal hazard (TT)	Temperature of smoke layer above 50 °C and bottom of smoke layer less than 1.5 m above floor
Class II conditions	Temperature of smoke layer above 100 °C
Class III conditions	Temperature of smoke layer above 160 °C
Class IV conditions	Temperature of smoke layer above 260 °C
Class V conditions	Temperature of smoke layer above 500 °C
Flashover (FO)	Temperature of smoke layer above 500 °C

These warnings are presented by SDFM in tabular form (Tables 5 & 6). The total heat release rate (HRR) is included in the tables. SDFM uses the ceiling temperature measurement in the fire room to calculate the convective HRR and then total HRR is estimated assuming a radiative fraction of 0.37. In addition, the SDFM predicted average smoke layer temperature for each room is shown graphically superimposed on the temperature measurements obtained from the thermocouple array in that room (Figures 4-8, 10-13). Additional graphs (Figures 9 & 14) present the predicted layer height from the SDFM as a function of time.

Fire Room 3

The fire room door opened at the middle of a long corridor as shown in the Figure 1. At one end of the corridor, it joined a second corridor at a right angle to it, while at the other end it connected to a second corridor in a "T" connection. Thermocouple arrays were located in the following positions (Figure 1):
- H inside fire room 3 near the door to the corridor
- G centered in the corridor outside the door of fire room 3
- C centered in the corridor outside the door of fire room 4
- B centered in the right angle section of the two corridors
- J centered in the "T" of the other corridor connection
- E and F located across from each other 1.8 m (6 ft) down the corridor from fire room 3 and 0.3 m (1 ft) off the wall

The door to fire room 4 was closed during the experiment. The SDFM was set to respond once a fire reached a level of 4 kW. Since the multiroom calculation of the SDFM is calculated using a zone model approximation, it was decided to break the corridor up into a series of virtual rooms, each room being equal to the width of the adjacent room as shown in Figure 2. This was done since a zone model provides the average temperature and smoke layer height for each room based on the conservation of mass and energy and averaging over long corridors will not capture the temperatures at the corridor extremes.

The corridor consisted of rooms 2 to 9 with rooms 5, 8, and 9 vented to the outside.

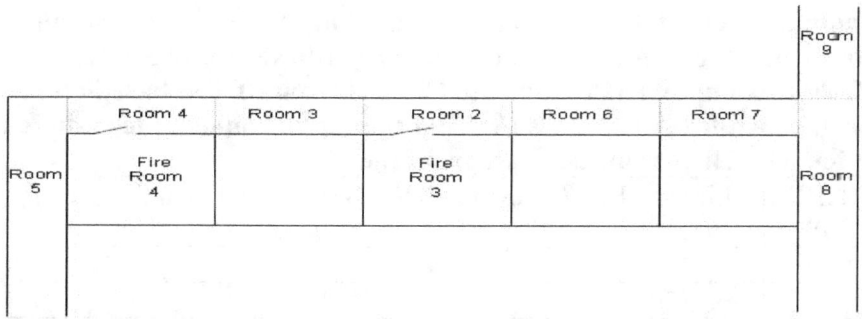

Figure 3. Modeling layout for the fire tests. Dashed lines are virtual boundaries used to simulate the corridor as a series of rooms.

Table 5 Fire Room 3 conditions as defined in Table 4. The numbers in parenthesis in the columns refer to Classes based on the top ceiling thermocouple (CMHA) in the room while the numbers without the parenthesis are predictions from the SDFM. For example, II(IV) indicate Class II from SDFM based on layer temperature while (IV) is Class IV from the top thermocouple assumed to be in the ceiling jet. The HRR is an estimate from the SDFM.

Time s	HRR kW	Fire Rm 3	Rm 2	Rm 3	Rm 4	Rm 5	Rm 6	Rm 7	Rm 8	Rm 9
0	4									
7	16									
14	97									
21	170	VL								
28	470									
35	860	TT(II)								
42	1200	(III)								
49	1400	II(IV)	VL							
56	1700	III	II(II)							
63	1300			VL		VL				
70	1500	IV	III(III)		VL		VL			
77	1600			II		II				
84	1900	(V)								
91	1700		IV(IV)	III		III				
98	1800			II (II)			II			
105	1700				VL					
119	1500				(II)					
126	1500				III					
133	1600							III	(II)	
140	1700				III				II	
154	1400					II				

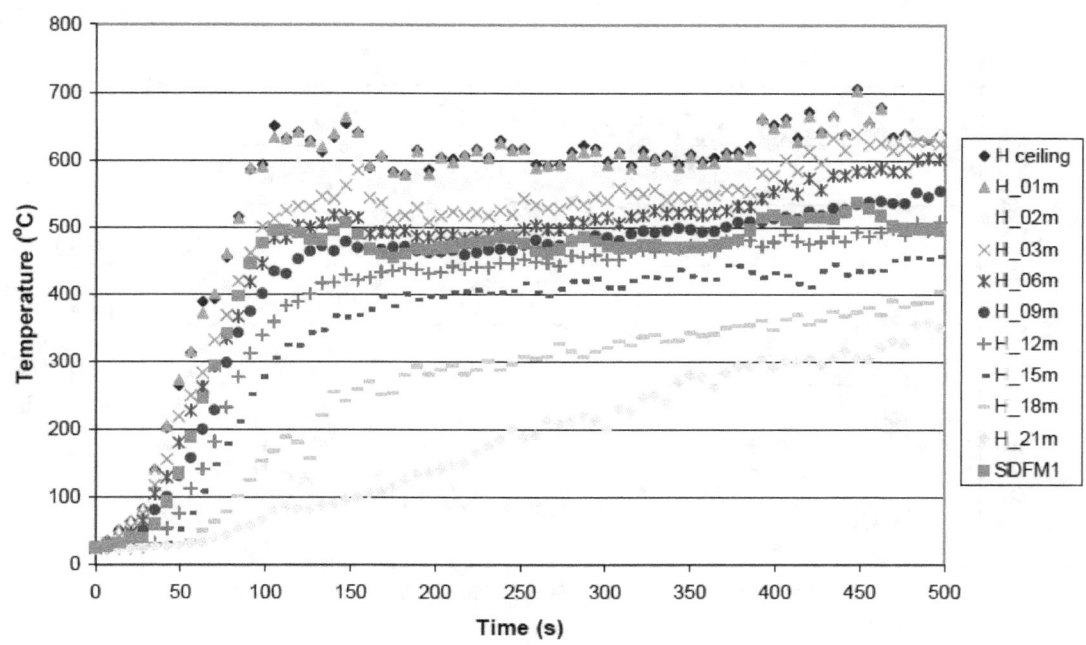

Figure 4. Fire Room 3. The squares are the prediction of the average layer temperature using SDFM where SDFM1 designates the fire room predictions. The rest of the plots are thermocouple data at the designated location below the ceiling.

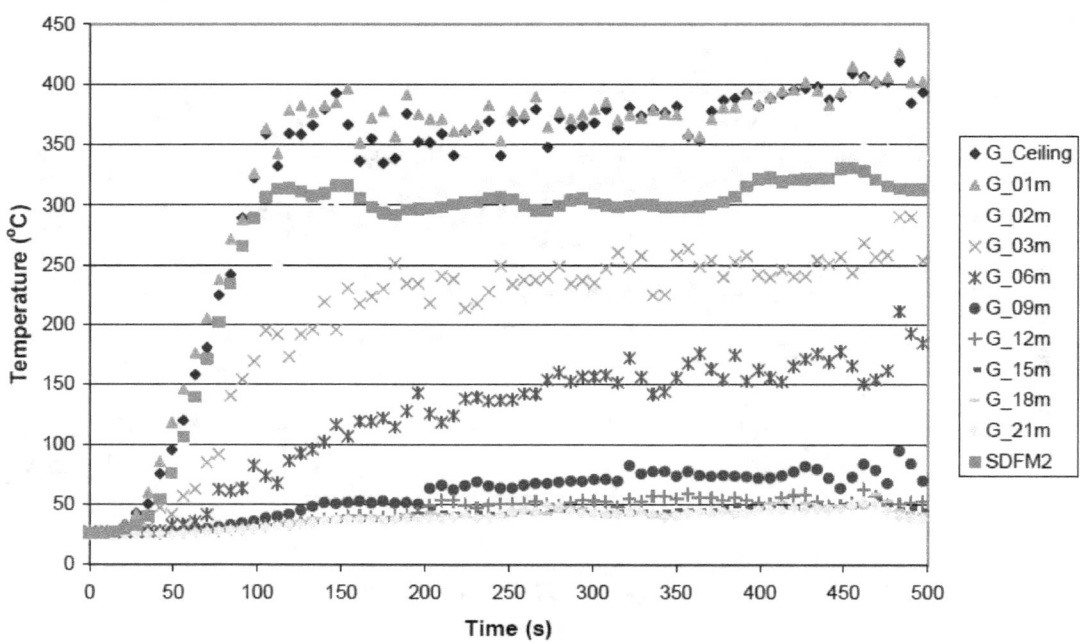

Figure 5. Hallway labeled Room 2. The squares are the prediction of the average layer temperature using SDFM where SDFM2 designates room 2 predictions. The rest of the plots are thermocouple data at the designated location below the ceiling.

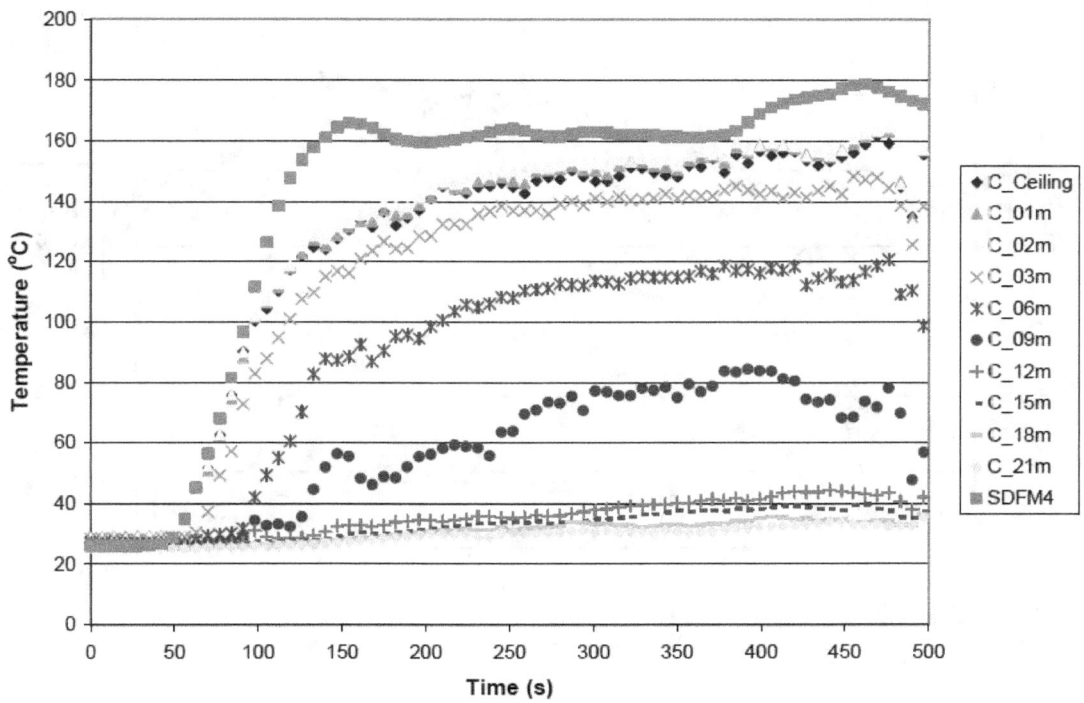

Figure 6. Hallway labeled Room 4. The squares are the prediction of the average layer temperature using SDFM where SDFM4 designates room 4 predictions. The rest of the plots are thermocouple data at the designated location below the ceiling.

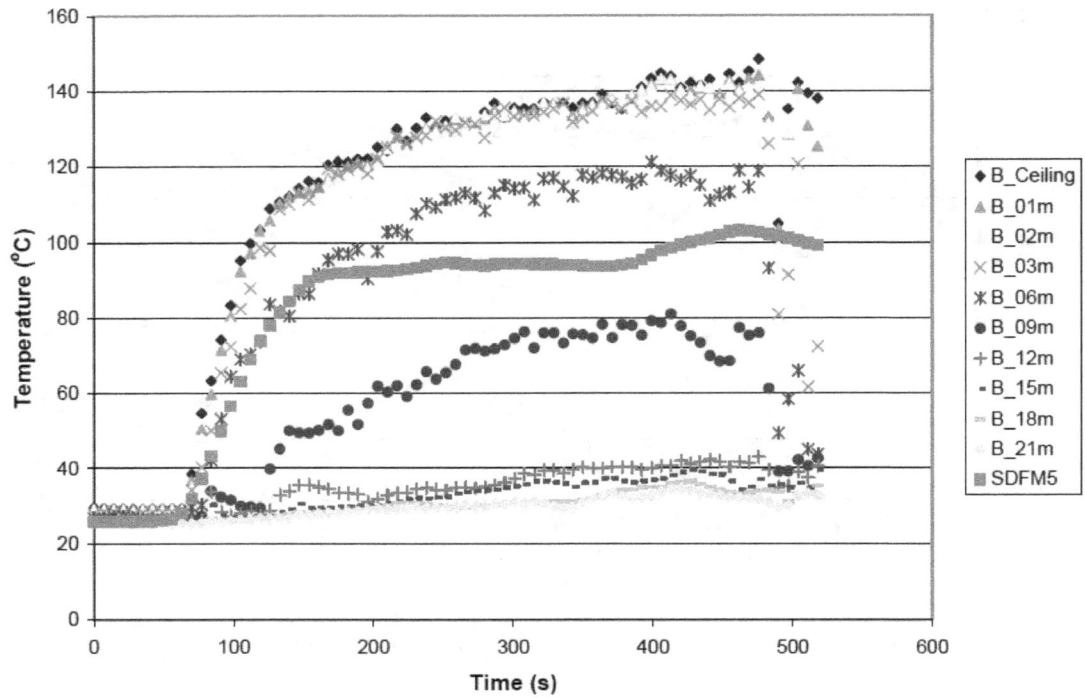

Figure 7. Hallway labeled Room 5. The squares are the prediction of the average layer temperature using SDFM where SDFM5 designates predictions in room 5. The rest of the plots are thermocouple data at the designated location below the ceiling.

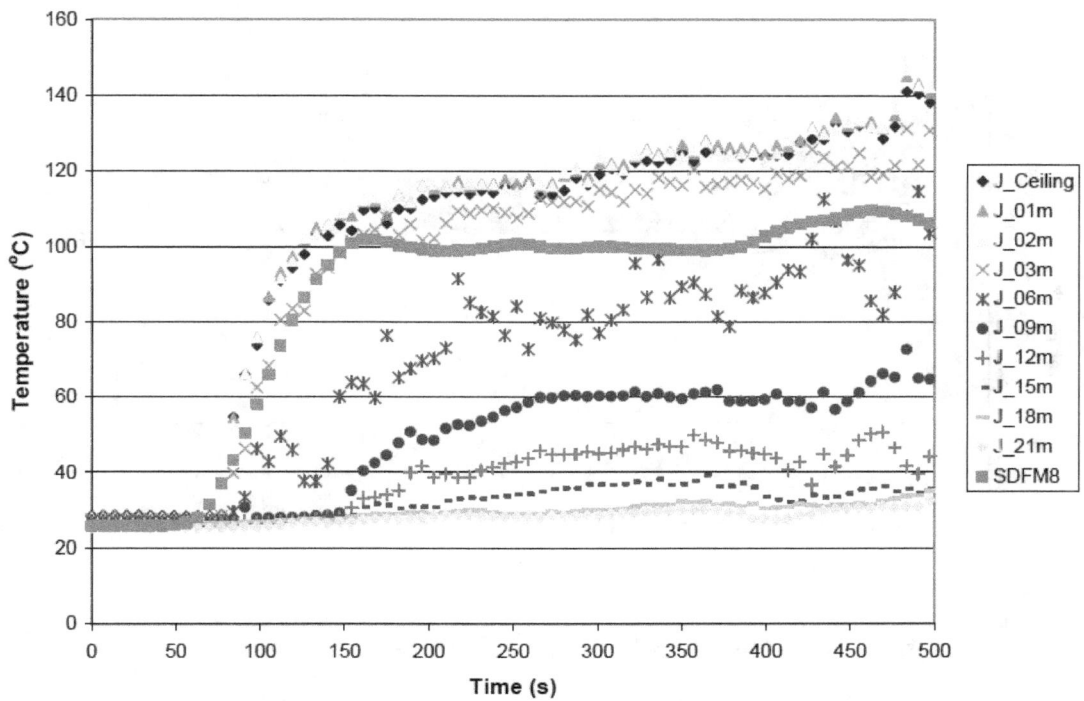

Figure 8. Hallway labeled room 8. The squares are the prediction of the average layer temperature using SDFM where SDFM8 designates predictions in room 8. The rest of the plots are thermocouple data at the designated location below the ceiling.

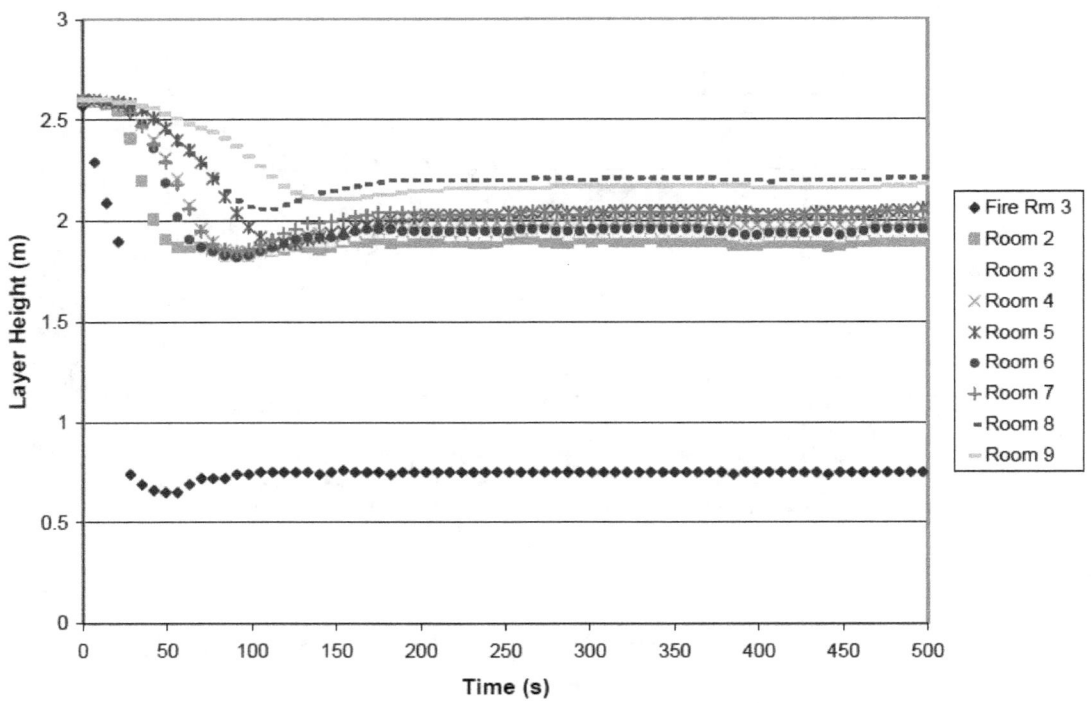

Figure 9. Layer Height measured from the floor as predicted by SDFM.

Fire Room 4

The fire room door opened at the one end of a long corridor as shown in the Figure 1 where it joined a second corridor at a right angle to it while at the other end; it connected to a second corridor in a "T" connection. Thermocouple arrays were located in the following positions (Figure1):
- D inside fire room 4 near the door to the corridor
- G centered in the corridor outside the door of fire room 3
- C centered in the corridor outside the door of fire room 4
- B centered in the right angle section of the two corridors
- J centered in the "T" of the other corridor connection
- E and F located across from each other 1.8 m (6 ft) down the corridor from fire room 3 and 0.3 m (1 ft) off the wall

The door to fire room 3 was closed during the experiment. The SDFM was set to respond once a fire reached a level of 4 kW. Since the multiroom calculation of the SDFM is done using a zone model approximation, it was decided to break the corridor up into a series of rooms, each room being equal to the width of the adjacent room as shown in Figure 3. This was done since a zone model provides the average temperature and smoke layer height for each room based on the conservation of mass and energy and averaging over long corridors will provide unrealistic temperatures at the corridor extremes. The corridor consisted of rooms 2 to 9 with rooms 5, 8, and 9 vented to the outside.

Table 6 Fire Room 4 conditions as defined in Table 4. The numbers in parenthesis in the columns refer to Classes based on the top ceiling thermocouple (CMHA) in the room while the numbers without the parenthesis are predictions from the SDFM. For example, III(IV) indicate Class III from SDFM based on layer temperature while (IV) is Class IV from the top thermocouple assumed to be in the ceiling jet. The HRR is an estimate from the SDFM.

Time s	HRR kW	Fire Rm 4	Rm 2	Rm 3	Rm 4	Rm 5	Rm 6	Rm 7	Rm 8	Rm 9
0	6									
7	12									
14	92									
21	100	VL								
28	135									
35	200									
42	280									
49	400	TT(II)								
56	700									
63	1000	(III)			VL					
70	1100	II								
77	800	III(IV)			(II)					
84	1000			VL	II					
91	1100				(III)					
98	1500									
105	1700	IV	VL	II	III	VL(II)				
112	1900						VL			
119	1700	(V)			(IV)	II		VL		
126	1600		(II)	III	IV					
133	1900		II							
140	1700	FO				(III)				
147	1700					III				
168	1300	TT								
231	1600	FO								
245	1600						II			
301	1500	TT								VL
308	1600	FO								
336	1600								(II)	

Table 4 Fire Room 4; Conditions include: Visibility Limited (VL), Toxic/Thermal hazard (TT) and Flash Over (FO). The numbers in parenthesis in the columns refer to Classes based on the ceiling thermocouple in the room (CMHA) while the number without the parenthesis is the predicted Class using the SDFM. For example, III(IV)

indicate Class III from SDFM based on layer temperature while (IV) is Class IV from the top thermocouple in the ceiling jet. The HRR is an estimate from the SDFM.

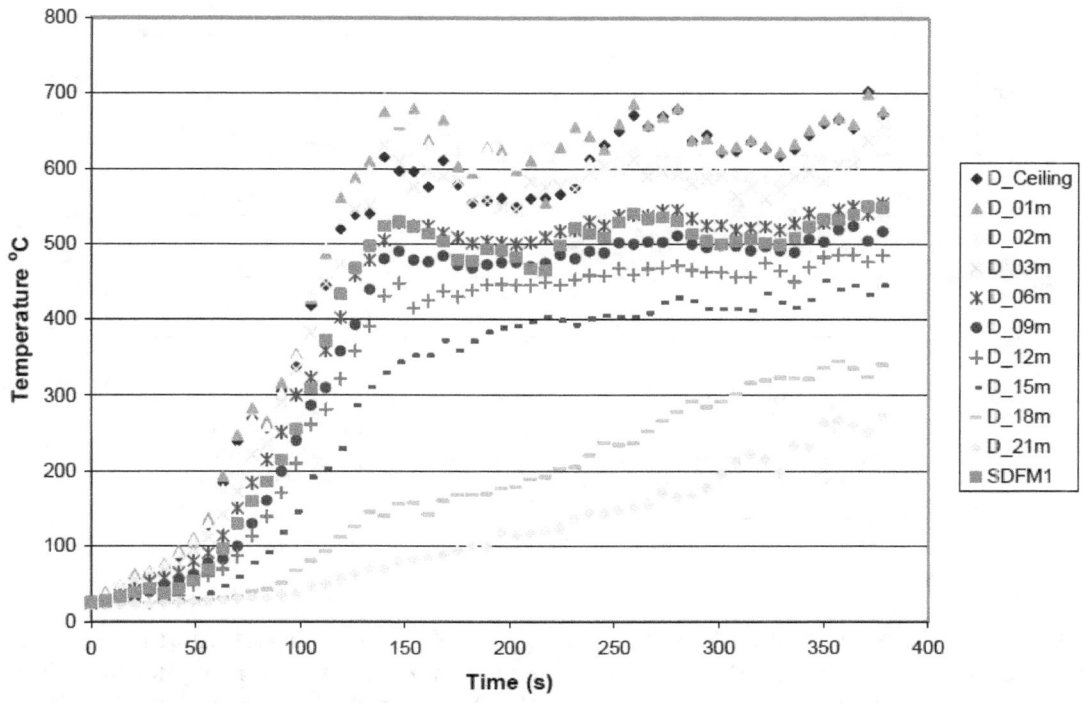

Figure 10. Fire Room 4. The squares are the prediction of the average layer temperature using SDFM where SDFM1 stands for fire room. The rest of the plots are thermocouple data at the designated location below the ceiling.

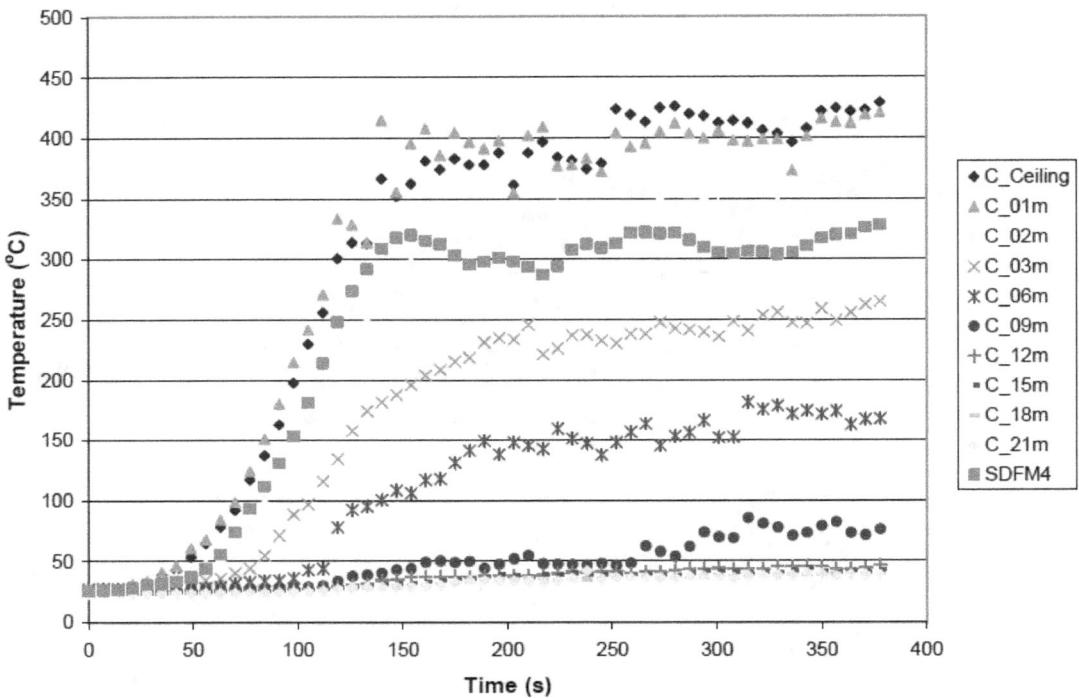

Figure 11. Hallway labeled Room 4. The squares are the prediction of the average layer temperature using SDFM where SDFM4 stands for room 4 prediction. The rest of the plots are thermocouple data at the designated location below the ceiling.

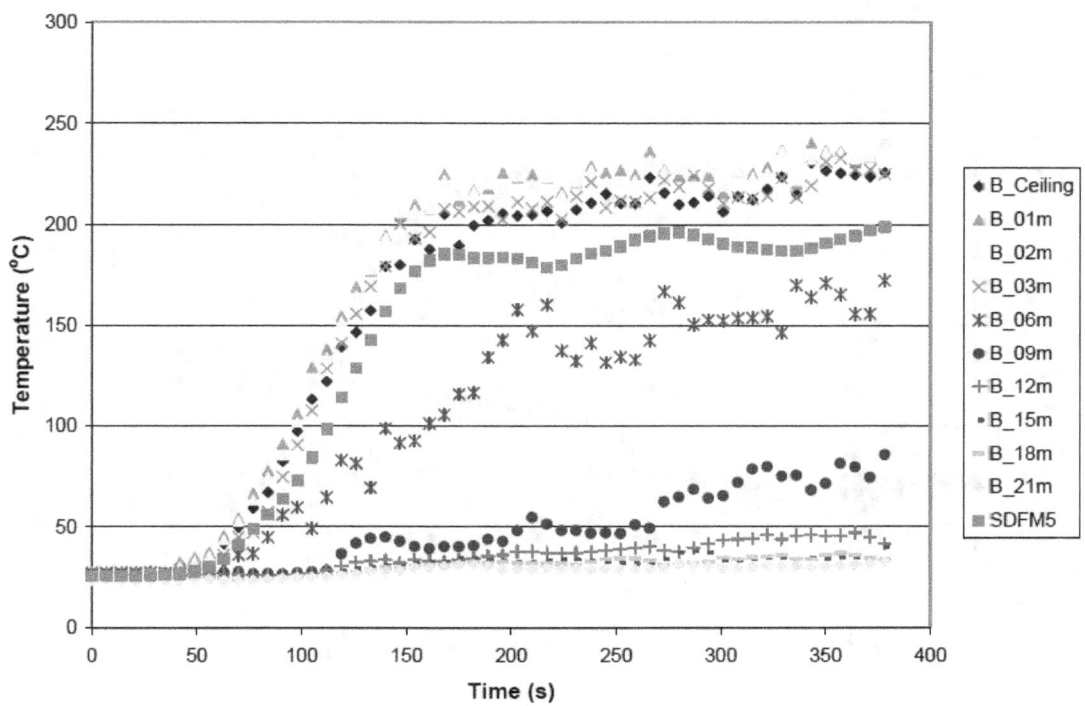

Figure 12. Hallway labeled Room 5. The squares are the prediction of the average layer temperature using SDFM where SDFM5 stands for room 5 prediction. The rest of the plots are thermocouple data at the designated location below the ceiling.

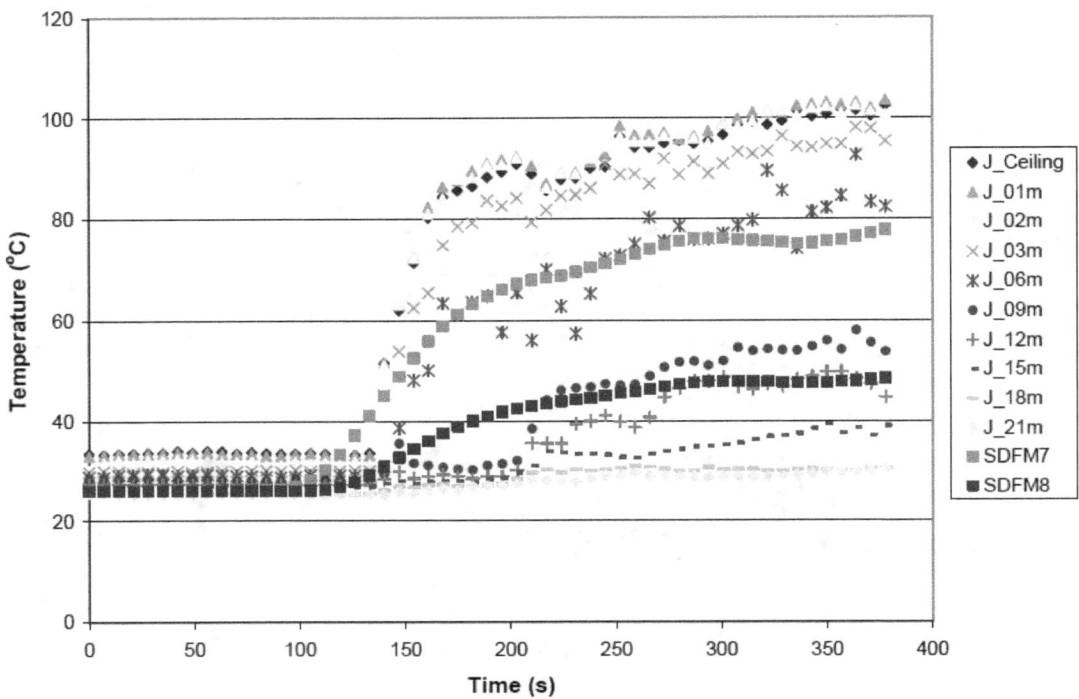

Figure 13. Hallway labeled Room 7. The squares are the prediction of the average layer temperature using SDFM where SDFM7 and SDFM8 stand for predictions for room 7 and 8. The rest of the plots are thermocouple data at the designated location below the ceiling.

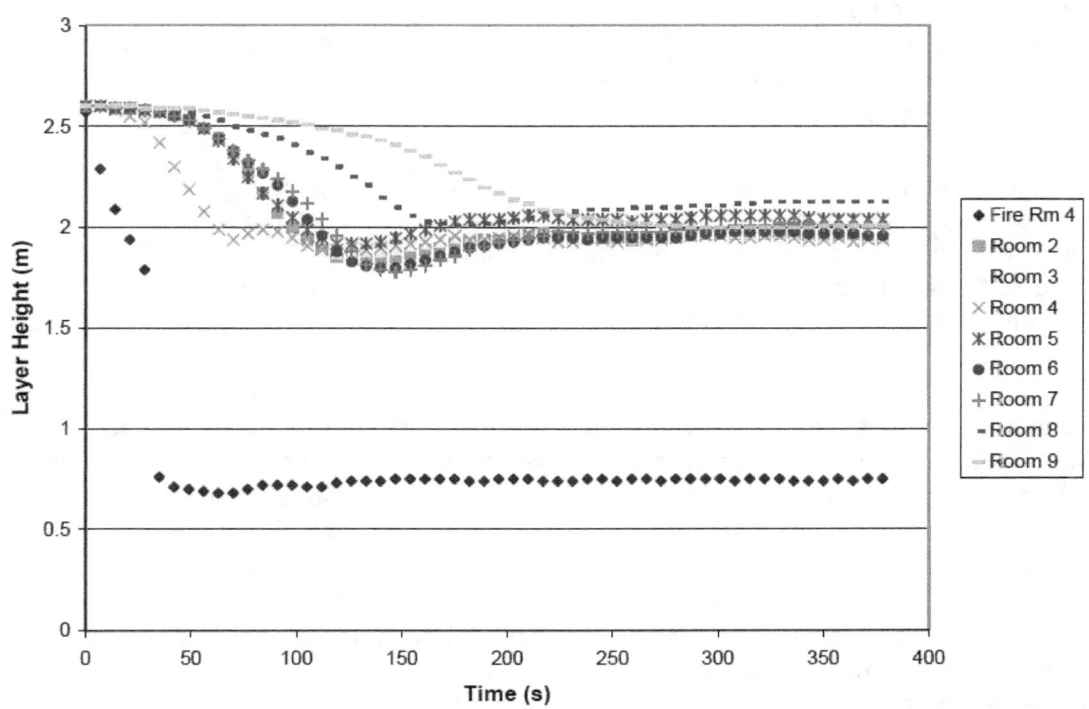

Figure 14. Layer Height measured from the floor as predicted by SDFM.

Discussion

The fire sizes for the two experiments were similar but the locations were substantially different with fire room 3 (FR3) feeding smoke into the central portion of the corridor while fire room 4 (FR4) fed smoke into one end of the corridor. The fire in FR3 developed more rapidly than the fire in FR4 with both rooms reaching flashover temperatures based on thermocouple readings at the ceiling but only FR4 reached the flashover criteria using the SDFM smoke layer prediction. The SDFM calculates an average smoke layer temperature which will be lower than the temperature in the ceiling jet and it is this temperature that is used to determine flashover by the model. The layer temperatures of both fire rooms were roughly 500 °C with FR4 being just hot enough to trigger the FO warning.

The graphical comparisons in Figures 4 thru 8 and Figures 10 thru 13 provide an instructive view of what a predicted smoke layer temperature (LT) indicates. For the fire rooms (Figures 4 and 10), the LT represents an average temperature roughly 20 % cooler than the very hot temperature of the ceiling jet. While there is a definite decrease in temperature as floor level is approached, the LT represents the environment that a standing emergency responder would be immersed.

The LT in the hallway provided a useful representation of conditions that emergency responders would encounter with only Figure 6 showing an over-prediction of temperature by roughly 25%. The agreement of SDFM predictions with the thermocouple arrays is quite satisfying considering that only the building geometry was known and the model used only the sensor signal in the fire room and the building ambient temperature to produce the temperature predictions.

Layer height represented another parameter that could be compared with the experiments. For both experiments, the layer height in the fire room was predicted by SDFM to be roughly 0.7 m, while in the hallway it ranged from just below to just above 2.0 m (see Figures 8 and 13). In examining the thermocouple data, in the hallway the temperatures transitioned to ambient at 1.8 m to 2.0 m, while in the fire room, the transition occurred at about 0.8 m.

An incident commander (IC) might use this information in the following manner. Assuming that the alarm is received at time 0s, the responders at the local fire station may require two minutes to get into gear and on the fire truck and another two minutes to reach the scene. Since it is a school, there may be three fire crews arriving within minutes of each other. Upon arrival on the scene and using the information delivered by the building fire system, the incident commander knows the location of the fire and that he has a substantial fire that has made the fire room and corridor outside the fire room too dangerous for the fire fighters to enter. The IC knows that the fire has not spread beyond the fire room and that search and rescue activities must avoid the hallway around the fire room. Since almost all the areas near the fire are visibility limited, the IC will need to use ventilation fans to remove the smoke and heat so that search and rescue activities can be conducted. The optimal way to attack the fire may be from the outside courtyard adjacent to the fire room; however, since the fire has been burning at a constant Heat Release Rate (HRR), it may be ventilation limited and a backdraft condition could be developing.

Since it is assumed that there are three teams available, a possible strategy might be to assign one team to set up a hose line for attack, a second team to do search and rescue in parts of the building judged safe for this activity while the third team would vent the fire room and set up ventilation fans at the appropriate doorway to remove smoke from the corridor in preparation to attack the fire. The electronic information from the building compliments and extends the information obtained as the IC does a visual size-up of the building fire. The electronic data provides the IC with insight about the fire that may not be evident from an external size-up. A visual size-up is still important to determine conditions outside the building as well as locate victims who may have escaped the building but are impaired from smoke inhalation.

It is interesting to compare the effectiveness of the two different types of DSS. In one case, the temperatures from the top thermocouples of the thermocouple arrays are used in conjunction with the temperature (see table 4) in order to provide a thermal hazard map for fire fighters. In the second case, the SDFM is used to provide information concerning fire size, visibility and thermal hazards. It should be noted that for most instances, the thermocouple hazard predictions using just the top thermocouples agreed with the SDFM predictions within a time span of 21 s suggesting that both methods are useful. The

SDFM provides fire size and visibility predictions as well, which can make this system more valuable.

Heat release rates (HRR) were not measured for these experiments, so no direct comparison could be made to validate the accuracy of the method used by SDFM. Indirectly, the agreement between the LT predicted by SDFM and the measured temperature profile from the thermocouple arrays suggest that the HRR estimates were usable. The convective HRR is used by the SDFM to predict smoke temperature and smoke layer height in each room of the building. A radiation fraction of 0.37 was used to convert the calculated convective HRR to a total HRR. For large fires, the radiation fraction could be closer to 0.2 which would reduce the total HRR to about 80 % of the estimated value[9].

Uncertainty Analysis

The uncertainty analysis for the experiments is taken from reference 3. "There are different components of uncertainty in the length, differential pressure and gas temperature data reported here. Uncertainties are grouped into two categories according to the method used to estimate them. Type A uncertainties are those which are evaluated by statistical methods, and Type B are those which are evaluated by other means [11]. Type B analysis of systematic uncertainties involves estimating the upper (+ a) and lower (- a) limits for the quantity in question such that the probability that the value would be in the interval (± a) is essentially 100 %. After estimating uncertainties by either Type A or B analysis, the uncertainties are combined in quadrature to yield the combined standard uncertainty. Multiplying the combined standard uncertainty by a coverage factor of two results in the expanded uncertainty which corresponds to a 95 % confidence interval (2σ)." For length measurements, the estimated expanded uncertainty is ±6% while the temperature measurements have an expanded uncertainty of ±15%.

Conclusion

It has been demonstrated that using temperatures from ceiling mounted heat detectors either directly or processing the values using a decision support system such as the SDFM can provide incident command with an assessment of thermal conditions in a building. While the SDFM contains algorithms that calculate smoke and gas concentration throughout a building, the instrumentation used with these experiments precluded testing these algorithms. Additional testing of these methods using realistic fires in full-size buildings is necessary to provide the validation required for use by the fire service. The decision support methodology discussed in the paper can operate in real-time and the information developed may be displayed on a laptop computer in a fire truck.

References

1 National Electrical Manufacturers Association SB 30, Fire Service Annunciator and Interface, Standards and Guideline Publications (2005).

2 W. D. Davis and G. P. Forney; Sensor-Driven Fire Model Version 1.1, NISTIR 6705, (2001) pp. 1-37.

3 Stephen Kerber and Daniel Madrzykowski; Evaluating Positive Pressure Ventilation in Large Structures: School Pressure and Fire Experiments, NIST Technical Note 1498 (2008) pp.1-359.

4 Davis, W. D., Holmberg David, Reneke, Paul, Brassell, Lori, and Vettori, Robert; Demonstration of Real-Time Tactical Decision aid Displays, NISTIR 7437, (2007) pp. 1-25.

5 Peacock, R. D., Jones, W. W., Bukowski, R. W.., and Forney, C. L.; Technical Reference Guide for the HAZARD I Fire Hazard Assessment Method Version 1.1, NIST Handbook 146, Vol. II, June 1991 pp. 1- 271.

6 Peacock, R. D.; Reneke, P. A.; Bukowski, R. W.; Babrauskas, V.; Defining Flashover for Fire Hazard Calculations, Fire Safety Journal, Vol. 32, No. 4, 331-345, June 1999.

7 McCaffrey, B. J., Purely Buoyant Diffusion Flames: Some Experimental Results. Final Report, NIST IR 1910 (1979) pp. 1-49.

8 Donnelly, M. K., Davis, W. D., Lawson, J. R., and Selepak, M. J.; Thermal Environment for Electronic Equipment Used by First Responders, NIST TN 1474 (2006) pp. 1 - 36.

9 Yang, J. C., Hamins, A., and Kashiwagi, T., Estimate of the Effect of Scale on Radiative Heat Loss Fraction and Combustion Efficiency, Combustion Science and Technology, 96, (1994) pp. 183-188.

www.ingramcontent.com/pod-product-compliance
Lightning Source LLC
Chambersburg PA
CBHW081811170526
45167CB00008B/3399